21st CENTURY INVENTIONS
EVERYDAY INVENTIONS

WORLD BOOK

www.worldbook.com

Co-published by agreement between Shi Tu Hui and World Book, Inc.

Shi Tu Hui
Room 1807, Block 1,
#3 West Dawang Road
Chaoyang District, Beijing 100025
P.R. China

World Book, Inc.
180 North LaSalle Street
Suite 900
Chicago, Illinois 60601
USA

© 2025. All rights reserved. This volume may not be reproduced in whole or in part in any form without prior written permission from the publisher.

WORLD BOOK and the GLOBE DEVICE are registered trademarks or trademarks of World Book, Inc.

Library of Congress Control Number: 2024943004

21st Century Inventions
ISBN: 978-0-7166-5360-8 (set, hard cover)

Everyday Inventions
ISBN: 978-0-7166-5363-9 (hard cover)
ISBN: 978-0-7166-5369-1 (e-book)
ISBN: 978-0-7166-5366-0 (soft cover)

WORLD BOOK STAFF

Editorial

Vice President
Tom Evans

Senior Manager, New Content
Jeff De La Rosa

Manager, New Product Development
Nicholas Kilzer

Content Creator
Elizabeth Huyck

Writers
William D. Adams
Lauren Kelliher
Fred Maxon
Jenna Neely

Proofreader
Nathalie Strassheim

Indexer
Nathaniel Lindstrom

Graphics and Design

Senior Visual Communications Designer
Melanie Bender

Digital Asset Specialist
Rosalia Bledsoe

Acknowledgments

Designer: Starletta Polster
Writer: Alex Woolf

CONTENTS

Adaptive clothing 4
Advanced sports analysis 6
Air fryer 10
Apps 12
Augmented reality 14
Bladeless fan 16
Bluetooth 18
Car charging stations 20
Consumer GPS 22
Consumer virtual reality headsets 24
Cosmic Crisp 26
Crowdfunding 28
Delivery robots 30
Domestic robots 32
E-readers 36
Emoji 38
Fitness trackers 40
Flexible touchscreens 42
Hoverboards 44
Image filters 46
Miniaturized cameras 48
N95 mask 50
Owlet Dream Sock 52
Personal safety alarms 54
Self-balancing scooter 56
Self-driving cars 58
Selfie stick 62
Smart refrigerators 64
Smartphones 66
Social media 68
Tablet computers 72
Video sharing 74
Virtual assistants 76
Index 78
Acknowledgments 80

Adaptive clothing:
Designs for every body

Putting on clothes should be quick and simple. But for some people with physical challenges, dressing can be frustrating. Clothing designers are responding with adaptive clothing.

Adaptive clothes and footwear accommodate different abilities and needs. Some have larger openings to fit a *prosthesis* (artificial limb). Instead of buttons or zippers, shirts and pants might fasten with magnets or Velcro, which can be used hands-free or with one hand. Some alterations allow people to receive care without having to remove clothing. Shirts that open along the arm or side give doctors easy access. These changes can help people with limited motion dress more easily.

A person who dresses while seated often needs slip-on pants or shoes. Adaptive footwear has magnetic laces or zips around the top of the shoe. These make it easier to put on shoes. Some companies sell shoes individually or in pairs of mixed sizes and widths. This helps accommodate differences in foot size or shape, or a prosthesis.

The American fashion designer Mindy Scheier founded the organization Runway of Dreams in 2013. She was inspired to develop fashionable and accessible clothing for her son, who has *muscular dystrophy*—a disease that affects movement, posture, and breathing. Scheier modified retail clothes to make them adaptive. She partnered with the clothing company Tommy Hilfiger to adapt its designs. In collaboration with Runway of Dreams, Tommy Hilfiger launched the first adaptive designer clothing line in 2016.

Adaptive fashion is about more than just looking good. Modified clothes help people to dress without needing assistance and thus give people confidence and dignity.

Wheelchair friendly

Wheelchair users can benefit from thoughtful clothing design. Jackets with side openings or zip-off sleeves are easier to get on and off. Adaptive fashion designers place fasteners where they will not rub against skin and avoid thick seams and tight fits. These features put pressure on joints and limbs for those with limited movement.

Express yourself

Following (or not following) trends and stylish dress are a big part of self-expression. Clothes should be functional, comfortable, high quality, and long-lasting. They should also match the wearer's style. When well-known brands create fashionable adaptive clothes, others follow suit. Adaptive clothing options then become more available to those who need them.

ADVANCED SPORTS ANALYSIS:
Crunching numbers to win

Whatever the sport, players and staff do everything they can to gain an edge over their opponents. One powerful new tool they use is sports analytics.

Sports analytics combines precise measurements from cameras and body sensors with statistics and algorithms (computer programs) to analyze player performance, game strategy, and team management. Today, all major professional leagues use video recording and wearable technology to review games and practices. These devices can precisely measure things like ball speed, throwing angles, and impact force. This information can help players and teams improve. The technology is even beginning to make its way into amateur leagues.

Imagine you're watching a soccer game. A player dribbles upfield, splits two defenders, and bends a powerful strike around a third and past the diving goalkeeper. Later in the game, another player clumsily bounces the ball right toward the goalkeeper. The goalie, expecting the ball to be struck harder, falls off balance, and the ball rolls into the net. Who made the better play? Everyone who watched the game would agree that the first player did, but the plays look the same on the scoresheet!

When is a player lucky and when is a player good? Did they benefit from good play by their teammates or poor play from their opponents? These questions are fun to argue about. But a lot is riding on the answers. In the past, even sports professionals often relied on instincts to form teams, coach players, or improve their play. But now they can check the data.

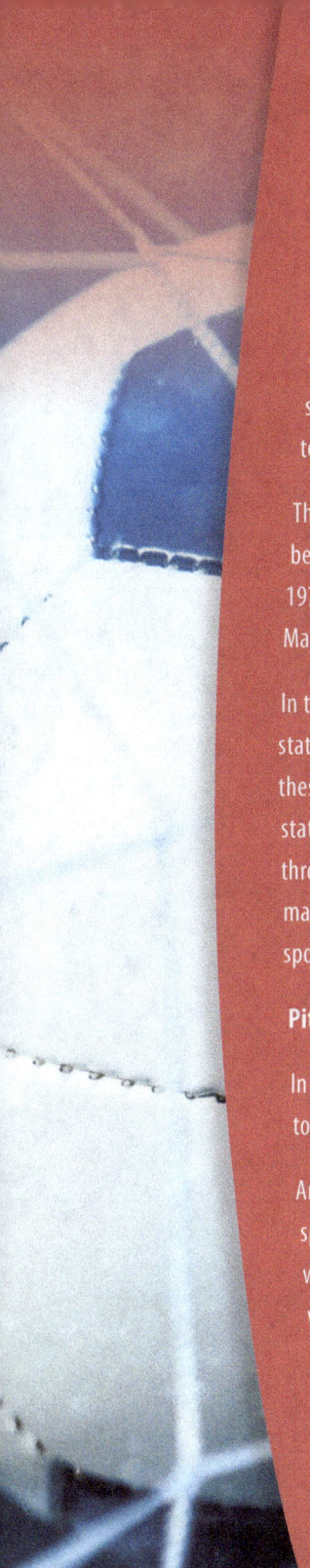

Baseball beginnings

Baseball was the first sport to make wide use of advanced analytics. Even before advanced play video, baseball teams liked to collect statistics on their players. Often this centered on two-part interactions: pitcher versus batter, fielder versus batted ball. However, the most common stats did not always capture how much a player helped his team win.

The American writer Bill James was one of the first people to notice this disconnect between common statistics and how good players were. Beginning in the late 1970's, James developed other statistics to better quantify player performance. But Major League Baseball (MLB) teams ignored James's ideas.

In the mid-1990's, the Oakland A's new general manager, Billy Beane, used James's statistics to identify good players that other teams undervalued. He could hire these players for less and build a competitive team on a shoestring budget. Beane's statistically driven A's enjoyed several winning seasons and playoff appearances through the 2000's. Within a decade, advanced statistical analysis became mainstream throughout baseball. The approach, dubbed "moneyball," spread to other sports as well.

Pitch-tracking and more

In 2006, a new three-camera video system, called PITCHf/x, allowed baseball teams to analyze pitches in great detail.

Analyzing the camera data revealed two key ingredients of successful pitches: speed and amount of spin. Faster pitches with more spin are harder to hit. This was already known, but pitch tracking showed just how much more valuable it was than any other factor. Coaches started training pitchers to throw as hard as possible. As pitch tracking technology improved, it allowed pitchers to work on spinning the ball as well.

In 2015, a new system called Statcast was installed in all Major League parks in the United States. Statcast gathered data on batted balls and fielders. Data on exit velocity and launch angle enabled batters to improve their swings to evade fielders. Defense, long the most difficult facet of baseball to measure, can now be quantified as well.

Virtual replay

Other sports involve more complex interactions than baseball. To analyze whole games, powerful computers can combine data from video tracking and wearable technology into a digitized version of how each game played out. By analyzing many digitized games, coaches can develop more optimal strategies or spot which players are most effective.

Catapult-ing forward

During practice, National Football League (NFL) players wear special vests designed by the Australian company Catapult. Each vest collects hundreds of data points per second, tracking the wearer's position on the field, speed, heart rate, and other information. Coaches use these data to evaluate the performance of players and particular plays. They can even estimate player fatigue and detect some injuries.

Catapult also makes a trackable football. A computer chip is sealed inside the ball during manufacturing. The ball can track spin rate, velocity, and other information.

The downsides of data

What happens when winning is boring? Not everyone likes the impact of data on games. In baseball, players are now encouraged to swing as hard as possible, hoping for a home run. Pitchers respond by trying to throw the ball past the batter. The result is more strikeouts and pitcher injuries. Some fans dislike this style of play. They want more bunting and base stealing, and longer outings by starting pitchers. In recent years, MLB has changed rules to try to return these elements to the game.

AIR FRYER:
Cooking up something new

Everyone loves the satisfying crunch of fried foods. But fried foods can also be greasy, messy, and high in fat. This fact made the air fryer a popular appliance in the 2010's and 2020's.

Foods are traditionally fried by dunking them in hot oil. The air fryer, on the other hand, uses little to no oil. Instead, it uses blowers to circulate hot air around the food. The hot, blowing air cooks food more quickly and evenly than an oven or stovetop. The food sits in a basket of wire *mesh* (netting). The holes in the mesh allow the hot air to evenly cook every surface of the food. The food comes out of the air fryer with a crisp texture.

Air-fried foods do still use fat, but less than their deep-fried counterparts. Doctors say people should be careful about the amount and kinds of fat in their food. An air fryer can help with that.

The Dutch technology company Philips introduced countertop air fryers in 2010. Over the next decade, people purchased tens of millions of them. Today, many companies sell their own models. Some kitchen ranges even come with a built-in air fryer function.

The air-frying revolution is not just about health. Frozen foods and leftovers—including pizza—can be soggy when reheated in a microwave. Air-fried leftovers are crisp and tasty by comparison. The lack of oil can also make cleanup much easier. In addition, air frying tends to be safer and easier than deep frying—especially for beginners.

Fry at home

Day-to-day plans changed for everyone during the COVID-19 pandemic. Researchers reported an increase in the number of meals prepared at home. The air fryer shot up in popularity and sales during this time.

Don't overload the air fryer

Most air fryers can hold about 2 to 3 quarts or liters of food. The amount depends on how much the mesh basket can hold without getting crowded. Crowding stops air from flowing properly.

APPS:
Programs aplenty

An app, short for *application software*, is a downloadable computer program designed for phones, tablets, or watches. Apps are what let you do all those fun things on your devices, including games, shopping, music, news, learning, and social media.

The first apps were installed on IBM's smartphone Simon in 1994. They included an address book, calculator, mail, notepad, and sketchpad. Today, there are millions of apps that do all sorts of things. Most are designed to run on either Apple's iOS system or Android OS. Each operating system has its own app store.

Some apps are free, while others cost money. Many apps have a free starter version, sometimes with advertisements that play in the app. Then users can pay to unlock more features or levels, or to not see ads.

Once downloaded, an app shows up as a small tile on your device's screen. When you open the app, it runs on the device's operating system. Many new apps take advantage of smartphone cameras, touchscreens, and location tracking to offer filters, touchscreen-based games, and interactive maps.

Computers also use apps. If you have ever used the email program or word processor downloaded on your computer, you have used a computer app!

Apps dominate screen time
Studies show that over 90 percent of computer screen time is spent on apps. That means users are not using web browsers to access programs.

Snacking snake

The Snake game, installed on Nokia phones in 1997, was the first mobile game app. It involved using the arrow buttons to move a snake toward food and away from the walls. As the snake grew bigger with each snack, it became more difficult to maneuver!

Augmented reality:
An enhanced world

Have you ever passed a building or sculpture and wanted to know more about it? Imagine you had a special pair of glasses that could display information as you walked by, or show a video of the artist explaining any piece of art you look at.

Such glasses are one example of *augmented reality* (AR). AR technology adds layers of images, sounds, or other information onto what you see in the real world. Augmented reality is most often used on smartphones or through special computerized glasses. These devices use cameras and other sensors to track their position and *orientation* (facing). A computer can then overlay information based on the view through the phone's camera or the glasses.

Augmented reality became a popular marketing tool in the 2010's. It enabled users to "try on" watches and jewelry from home by adding a virtual image of the item onto a live image of the user, captured through the device's camera. Face-tracking technology lets AR users try virtual makeup, hairstyles, and eyeglass frames. It can also add funny features such as cat ears that move with your face in video calls.

Museums have found many fun uses for augmented reality. The Museum of London released an AR app in 2010 that overlaid historical photographs onto modern views of London, England. Astronomy AR programs show star maps and other astronomical information when a user points their camera at the sky.

Foul ball

The AR company Hawk-Eye produces technology for spectators at Major League Baseball games. It overlays views of the game with ball-tracking and other game details. Even fans way up in the "nosebleed" seats can play umpire!

Automobile manufacturers have experimented with AR displays to assist drivers. These show hazard information and driving directions over views of the road. AR navigation on smartphones can help pedestrians find their gate at the airport or locate a particular restaurant.

I choose you!

The AR game Pokémon Go gathered 232 million users in 2016. The game enabled players to find and battle virtual Pokémon monsters in their real-life surroundings. Imagine spotting Pikachu in your yard or local park!

BLADELESS FAN:
A cooler way to stay cool

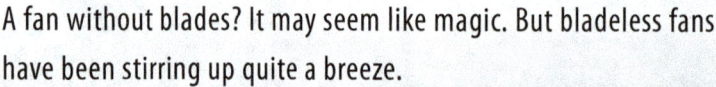

A fan without blades? It may seem like magic. But bladeless fans have been stirring up quite a breeze.

How can such a fan blow air? The "bladeless" fan actually does have blades, but they are small and hidden away in its base. As a result, bladeless fans can be safer and lighter than traditional fans. A clever design enables the bladeless fan to produce a steady, powerful stream of air.

When the fan is turned on, hidden blades in the base pull in air through a small vent. The air flows up into a circular or oval-shaped ring. A channel in the ring controls the airflow and pushes it out in the desired direction.

As the air pulled in by the fan swirls around the ring, it creates a vortex that pulls in more air from behind the fan in a process called *inducement*. As air leaves the fan, it also draws in surrounding air, in a process called *entrainment*. Inducement and entrainment give bladeless fans a mighty stream of air with only a small motor and blades.

Bladeless fans do not just look cool; they actually have several advantages over conventional bladed fans. Conventional fans chop up the air with their blades, creating turbulence and noise. Bladeless fans produce a much steadier breeze. They also use less electricity, making them cheaper to run and better for the environment. And with no exposed blades to collect dust, they are easier to clean.

History of an idea

The idea for the bladeless fan was first developed by the Japanese electronics manufacturer Toshiba in 1981. The design and technology were improved by the British inventor James Dyson, whose company released their version in 2009. Dyson called his design the Air Multiplier because it increased the flow of air a whopping 15 times.

Noise reduction

Early models of the bladeless fan were quite noisy, because of the motor in the base. Engineers introduced into the base a hollow chamber called a Helmholtz cavity to trap sound waves from the motor. As a result, bladeless fans are now much quieter.

Bluetooth:
Connected without wires

It may be hard to believe in our wireless world, but not long ago most electronic devices were connected by a tangle of wires. That was before Bluetooth wireless technology. Bluetooth technology is a way for devices to connect without cables. But how does it work?

Bluetooth uses short-range radio waves at frequencies between 2.4 GHz and 2.480 GHz. Devices use these radio waves to exchange information.

The first step in using bluetooth is to "pair" two devices. Say you wanted to use wireless Bluetooth headphones with your phone. First, the headphones send out a packet of data called an *advertisement*. This signal announces the device's presence to other nearby devices. This ping causes the headphones to appear on the phone's Bluetooth menu. If you select them for pairing, the two devices will start exchanging signals over Bluetooth.

Bluetooth devices ensure a secure connection by frequency hopping, jumping from channel to channel within the Bluetooth frequency at the same time. They do this to avoid other devices connected over Bluetooth.

Bluetooth technology was first developed in the 1990's by Jaap Haartson, a Swedish electrical engineer. Now it is found in all sorts of devices, from smart home appliances to gaming systems.

An unlikely origin

One early pioneer of the frequency hopping technique used by Bluetooth devices was the Austrian American actress and inventor Hedy Lamarr. During World War II (1939-1945), Lamarr teamed up with the piano player George Antheil. They invented a system to control torpedoes by radio without interference from enemy craft. Their method was inspired in part by the long strips of paper used in player pianos.

Computer networks, robots, and networked appliances sometimes use a special form of Bluetooth called Low Energy. This is designed to communicate basic information like distance, speed, and direction of travel.

What's in a name?

Bluetooth gets its unusual name from the early Scandinavian ruler Harald Bluetooth. He united Norway and Denmark in the year 958 and was famed as a great communicator. Harald had a dead tooth that earned him the nickname Bluetooth. The Bluetooth logo is made up of his initials, H and B, written in runes, the writing system of his time.

CAR CHARGING STATIONS:
Electrifying!

Electric and hybrid cars and trucks save energy and cut down emissions. They are becoming more popular every year. These vehicles run on rechargeable batteries. But where do they recharge?

Most electric cars can travel 100 to 300 miles (160 to 500 kilometers) on a single charge. For everyday use, EV owners recharge their vehicles by plugging in at home or at local charging stations. But to travel long distances, they need public charging stations along the way, the same way gas cars need gas stations. Today, there are more than 130,000 charging stations in the United States. The U.S. Department of Energy plans to add 500,000 more to create a national network of EV charging stations.

EV manufacturers and other companies are working to make charging stations more widely available. They are placing charging stations in shopping malls, airports, and popular restaurant chains. The new charging stations are built so that a wide range of electric vehicles can use them.

A company called ChargePoint created the world's largest network of EV chargers. Drivers can find the nearest ChargePoint station and pay for their charge using a smartphone app. Another company called Red E Charge has partnered with the sandwich chain Subway to install EV chargers at its restaurants. Each Subway Oasis plans to have EV chargers, a park for recreation, and Wi-Fi access.

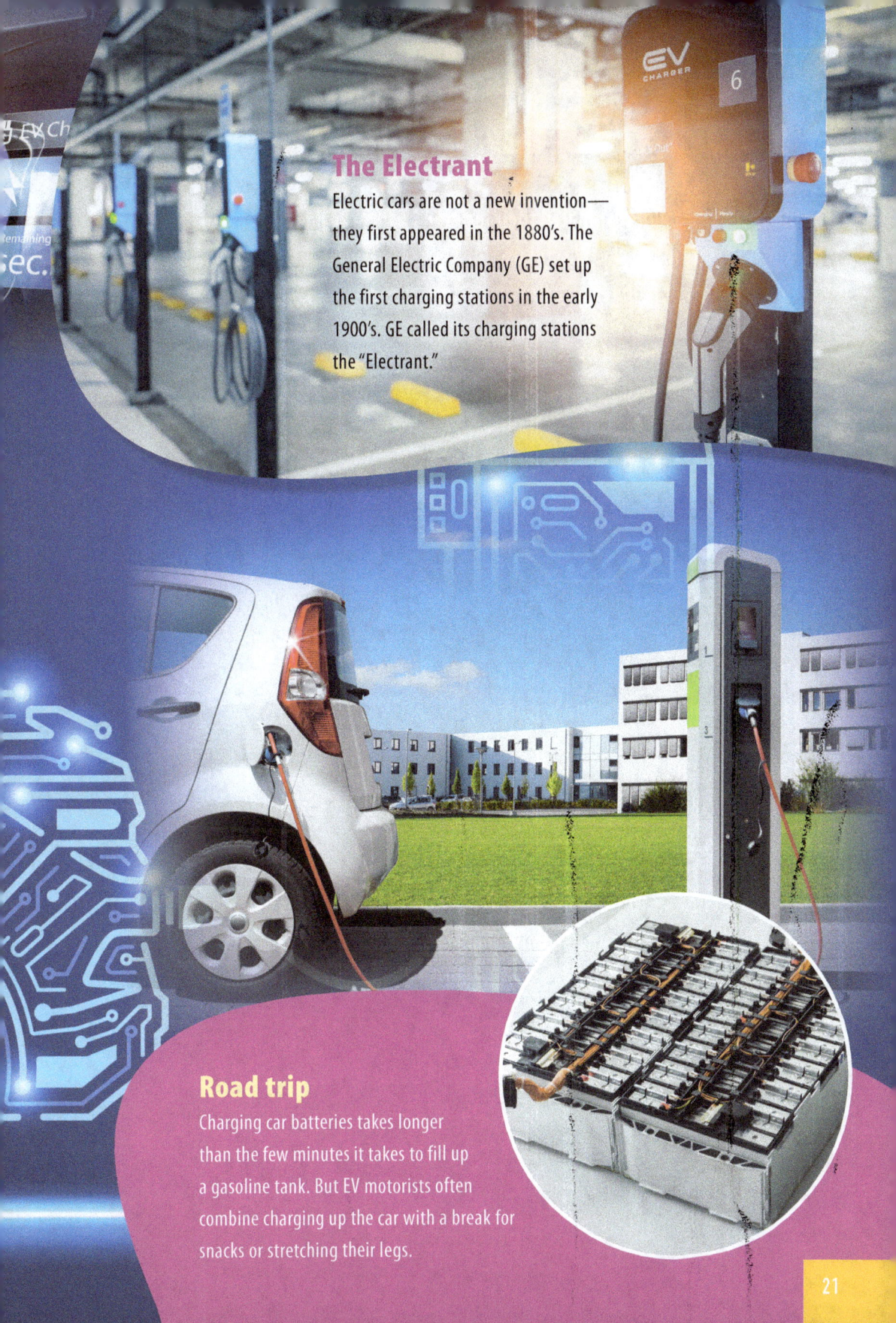

The Electrant

Electric cars are not a new invention—they first appeared in the 1880's. The General Electric Company (GE) set up the first charging stations in the early 1900's. GE called its charging stations the "Electrant."

Road trip

Charging car batteries takes longer than the few minutes it takes to fill up a gasoline tank. But EV motorists often combine charging up the car with a break for snacks or stretching their legs.

CONSUMER GPS

In ancient times, sailors estimated their location by measuring the angle of the sun and stars in the sky. We still use the sky to find our way around today, though it no longer requires an astrolabe.

The Global Positioning System (GPS) uses 31 satellites evenly spaced in orbit 12,600 miles (20,278 kilometers) above Earth to figure out your location. These satellites constantly send out time signals to GPS receivers in cars and cell phones. The difference between the time announced in the signal and the time it arrives tells the receiver how long it took to travel, and so how far away the satellite is.

Receivers use signals from three or four satellites to determine your location in a process known as trilateration. Imagine drawing a circle around each satellite with a radius of its distance from the receiver. Those circles will all meet at just one point—where you are.

Aircraft use Ground-Based Augmentation Systems (GBAS) for even better accuracy. GBAS looks for fixed points on Earth in addition to satellite data. This reduces errors and helps prevent collisions.

The United States Air Force and Department of Defense are responsible for maintaining the satellites and keeping them safe from any "space junk" that might hit them. As older satellites start to fail, newer models replace them with upgraded technology.

GPS was first introduced in 1973 as a part of the Department of Defense's Navigation Satellite Timing and Ranging (NAVSTAR) program. The first satellite was launched in 1978, and GPS was made available to airplanes in 1983. In 2000, President Bill Clinton opened access to accurate satellite GPS data for consumer use, paving the way for the familiar GPS systems we know today.

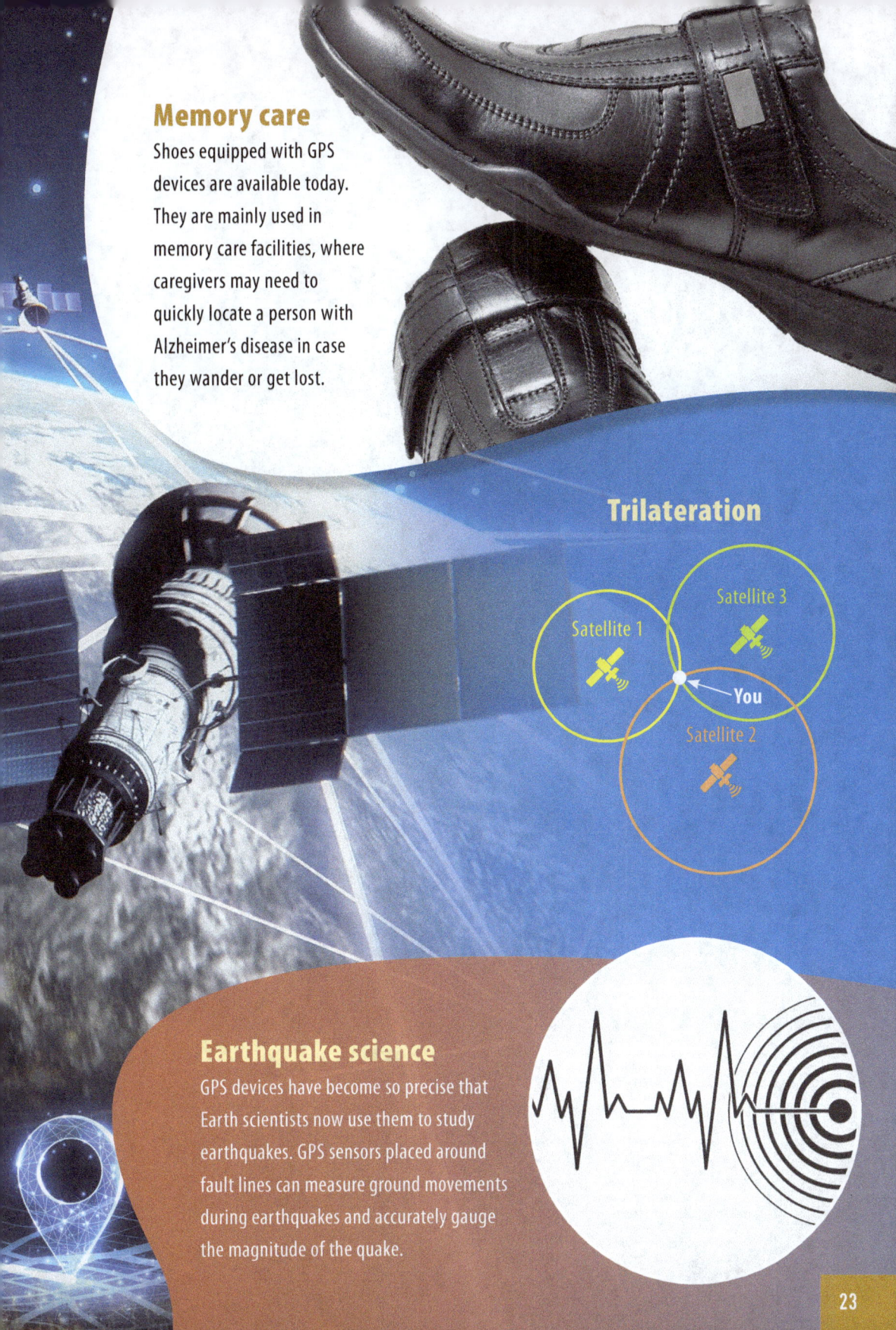

Memory care

Shoes equipped with GPS devices are available today. They are mainly used in memory care facilities, where caregivers may need to quickly locate a person with Alzheimer's disease in case they wander or get lost.

Trilateration

Satellite 1
Satellite 3
Satellite 2
You

Earthquake science

GPS devices have become so precise that Earth scientists now use them to study earthquakes. GPS sensors placed around fault lines can measure ground movements during earthquakes and accurately gauge the magnitude of the quake.

23

Consumer virtual reality headsets:

A tool to take you places

Imagine floating in outer space, swimming in a coral reef, hiking Mount Kilimanjaro, or attending a concert—all from your living room! Such trips are possible with *virtual reality* (VR), an artificial, three-dimensional computer environment.

VR is typically experienced by putting on a headset that acts like a wrap-around screen—with full sound and in 3D. Motion sensors track the movement of your head to let you look around the virtual world as if it were all around you. When you move your head to look up, the display shows you what is up in the virtual world. Sounds also shift between ears and change as you move, just like in the real world.

VR has been used in industry settings since 1970. The first consumer VR headset was released in 1991, but they did not get popular until the mid-2010's, when advances in technology made much better headsets possible. A company called Oculus released a popular VR system, the Rift, in 2016.

VR headsets can also be paired with VR gloves, controllers, and other accessories to add touch (haptics) or other senses to virtual worlds. For example, the gloves might vibrate if you get hit by an enemy in the game. Ouch!

VR headsets are not only used for gaming. You can also put one on to watch movies, go to concerts, learn how to play instruments, tour travel sites, try sports, meditate, or build 3D models. VR can also be used to shop online in a digital store you can interact with in real time. What an adventure!

VR sickness

VR makes some users feel dizzy or nauseous, or gives them headaches. This is caused by conflicting brain signals from the eyes and the body about whether you are moving, similar to motion sickness.

Face your fears

VR has been used to treat people with extreme anxiety and post-traumatic stress disorder (PTSD). By facing fears of spiders, heights, public speaking, and other triggers in the safety of VR, users can overcome them.

Cosmic Crisp:
The perfect apple

Science has improved many areas of our lives. But can scientists really make a better apple? A new variety of the fruit made its debut in 2019.

What makes the perfect apple? Many apple fans look for a firm and crisp texture. The fruit's taste should be sweet and tart. Most of all, an apple should be juicy.

This particular perfect apple was 22 years in the making. The Cosmic Crisp has not been genetically modified (had its genes altered in a laboratory). It was created through good old-fashioned cross-breeding.

The Cosmic Crisp is a *hybrid* (cross) of two popular apple varieties: the Honeycrisp and the Enterprise. The Honeycrisp was bred for flavor and snap, but does not keep very long in the refrigerator. The Enterprise was bred to resist disease and browning. The Cosmic Crisp combines the sweetness of the Honeycrisp with the hardiness of the Enterprise.

Bruce Barritt, a professor at Washington State University, first crossed the Honeycrisp and the Enterprise in 1997. In 2017, 300,000 Cosmic Crisp trees were planted. Over the next few years, growers planted more than 10 million trees.

The Cosmic Crisp was developed to grow in central Washington, the leading apple-growing region of the United States. Cosmic Crisp apples have high sugar and acid levels. This helps them to stay crisp and sweet longer. Fresh-picked Cosmic Crisp apples can be stored an entire year in the refrigerator!

Sweet origins

The Honeycrisp was created in the 1990's as a hybrid of Honeygold and Macoun apples. The Honeycrisp passed on to the Cosmic Crisp its juiciness, sweetness, and flavor. The Enterprise is a hybrid of a few varieties, including the Golden Delicious, McIntosh, and Rome Beauty—even the humble crab apple! The Enterprise gives the Cosmic Crisp its rich red color, disease resistance, and longevity. It also adds tartness, making the Cosmic Crisp great for baking.

The stars align

The new variety gets its "cosmic" name from the bright speckles on its skin, which are said to resemble stars.

Crowdfunding:
Kickstart your project with help from the crowd

Do you need money to start a business, recover from a disaster, or get your invention made? Why not let the internet crowd pitch in?

Crowdfunding is the idea of letting many individuals invest small amounts of money toward some specific goal. Some crowdfunding websites cater to charity, others to business or creative projects. GoFundMe, founded in 2010, was one of the first crowdfunding sites. Since then, it has raised over $15 billion from more than 100 million donors.

Typically, each crowdfunding project has a goal to raise a certain amount of money, anywhere from a few hundred to millions of dollars. Sharing campaigns on social media can reach millions of potential donors.

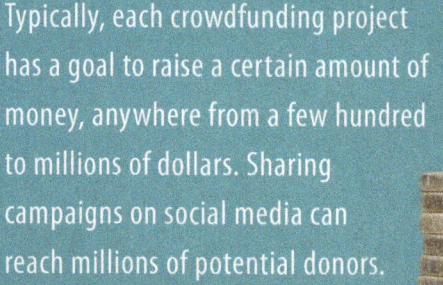

Launching an industry

One noteworthy Kickstarter campaign was launched by the company Oculus to build a virtual reality (VR) headset. The original goal was $250,000. The campaign ended up raising $2.4 million from donors eager to get their hands on the device. Two years later, the social media giant Facebook acquired Oculus for $2.3 billion.

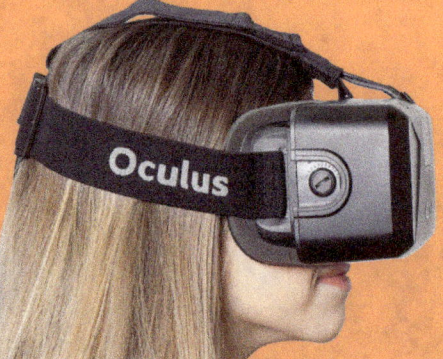

Crowdfunding websites take a portion of the total raised to cover their costs. These fees can range from 5 to 12 percent.

Often, funders are offered a perk or reward in return for funding a project. If you donate money to help make a motion picture, for example, you might get your name in the film's credits. A common reward for funding a new invention is receiving a sample of the manufactured item once it is made.

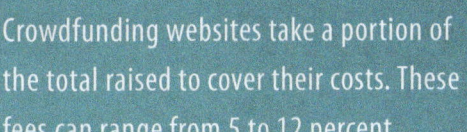

Feed the people

The charity campaign America's Food Fund became GoFundMe's highest earner. It was launched in 2020 and has raised over $45 million for the hungry. The campaign was started by the actor Leonardo DiCaprio and the philanthropist Laurene Powell Jobs.

Each crowdfunding platform works in its own way. Kickstarter only collects from donors and gives creators their money if the project reaches its donation goal. But on GoFundMe, the donated money is available immediately, even if the goal has not been met. GoFundMe is often used to raise money for personal or community causes. For example, a group might raise money to support victims of a natural disaster or to buy new equipment for a playground. When everyone pitches in, a little goes a long way.

DELIVERY ROBOTS:
Robot promise delivered

Delivering food or packages can be a dull job. Why not let robots do it?

Delivery robots are small, battery-powered, self-driving carts that deliver goods to your home. They're about the size of a small cooler and roll down sidewalks about as fast as a person walks. A small flag makes it easier for people to spot them when they are crossing the street. Each robot can hold roughly three bags of groceries or small parcels.

The robot's cargo box is kept safely locked during transit. If someone tries to tamper with the robot, it sounds a loud alarm. A phone code allows the customer to unlock the delivery when it gets to its destination.

Customers can track the location of the robot on their phone thanks to built-in GPS. Cameras and sensors help the robot find its way. If it runs into a problem, it alerts a human operator. The robot can remember

All-weather robots
Delivery robots have to be tough enough to deliver at night and in all kinds of weather. Current models are small, but larger delivery bots may come soon.

each time it overcomes an obstacle, so it learns a little more about its delivery route each time—making future deliveries smoother and safer.

Other robots have found work in hotels. Hotels use the Butlerbot to deliver extra bedding or room service food to guests. Like their outdoor counterparts, Butlerbots map their environment with sensors and cameras.

So if you ever see a cooler rolling down the sidewalk, step aside and let it pass. It may just be a robot on a delivery run.

Service, please!

Food-serving robots can wait tables in restaurants, taking orders and bringing customers their food.

DOMESTIC ROBOTS:
Helpers in the home

You're in your living room playing with your robot puppy. A robot vacuum cleaner senses his presence and steers around him. Outside, you can hear the gentle buzz of another robot mowing the lawn. If this sounds like the future, think again. Domestic robots are already here, and the more technology advances, the more they will be able to help us around the house.

Domestic robots assist with household chores. They can clean floors, cut the grass, clean swimming pools, and entertain children. They move around and perform their tasks automatically, with little or no need for human oversight.

The most common domestic robot today is the robotic vacuum cleaner, or robovac. This compact little robot, usually shaped

How robovacs work

The robovac uses sensors to navigate the floor of a room. The sensors can detect walls, obstacles, and steep drops, so the robovac doesn't bump into things or fall down steps. Special dirt sensors help it locate particularly dirty areas. If it gets stuck, it uses a synthesized voice to call for help.

The Roomba, a popular robovac, has a downward-facing infrared sensor and an upward-facing camera. Together, these allow the robot to map the floor area of a home so it can find its way back to its recharging station. An onboard computer remembers the map and improves it with each run. Other robovacs use laser guidance systems to create a floor plan.

like a disk, cleans carpets and hard floors. Some even mop. Robovacs can be programmed to suit individual needs, such as turning off at certain times or avoiding some areas. Some models can empty their own bags into a disposal unit.

Robomops

Robotic mops can automatically spray water onto a floor before moving over it. Advanced versions can detect and avoid carpeted areas.

33

When it's time to trim the lawn, a robotic lawnmower starts itself up and gets to work. How does it know what's lawn? Some rely on a boundary wire around the edge. Other models use cameras and sensors to follow a map. Some also contain rain sensors. If they detect rain, they automatically return to their recharging station.

Normally robotic mowers choose their own routes around the lawn, but they can also be steered with a phone app. When the mowing is done, the robot returns to its base and recharges.

Social robots provide entertainment and companionship for children, the elderly, or people who live alone. Some are humanoid in form, while others look like pets. Humanoid robots use a form of AI called natural language processing (NLP) to chat, respond to commands, play games, and tell jokes. Some are programmed with personalities. They might seem to be afraid of heights, for example, or exhibit tiredness, grumpiness, or joy. Other robots are designed for people with healthcare needs. They can remind patients to take medication, monitor vital signs, and, if necessary, alert healthcare professionals.

Furry robots

Robotic pets have built-in sensors that respond to stroking and cuddling. They can move and make sounds like their real-life animal counterparts. Pet a robo-cat's cheek and it will nuzzle its head into your hand. Pet its back or behind its head and it will purr happily and may even roll onto its back for a belly rub. Robotic pets are ideal for people who would love to have a pet, but can't keep a live animal.

Patrolling the home

Security robots watch over homes when residents are away. Their cameras and motion sensors can detect intruders. The robots can move around to check for trouble and send live video to mobile devices. Some have facial recognition software to help identify visitors. Security robots can also detect fires, floods, animals, and gas leaks, and alert emergency services.

E-readers:
An entire library in your hands

Have you ever wished you could bring all your favorite books on vacation, or find something new to read without leaving the couch? Now you can, with an *e-reader*. E-readers are devices that can download and display books in *electronic* (computerized) format. Such books are called *e-books*. (The *e* in both terms stands for *electronic*.)

E-readers use wireless connections to download book files from online stores or libraries. When you read a book on an e-reader, a tap on the touchscreen turns the page. An e-reader can hold thousands of books. And if you finish them all, you can download more over the internet.

Computers, phones, and tablets have screens that light up. Such screens can strain the eyes during long reading sessions. Instead, many e-readers use a gentler screen technology called *e-ink*. (You guessed it, *electronic* ink!) E-ink was invented in 1997, and the first e-ink reader appeared in 2004.

E-ink uses small electric charges to flip pixels in the reader screen to dark or light. The black-and-white screen can be read by *ambient* light—ordinary light falling on it.

The Kindle catches fire
The online retailer Amazon introduced its Kindle e-reader in 2007. The device sold out in just 5½ hours!

This makes it look more like a printed page. It is also easier to see in bright sunlight, for reading outside.

E-ink uses much less power than computer or phone screens, so e-reader batteries last a long time. Some e-readers include built-in lights for reading in the dark.

Many e-readers include tools to highlight, take notes, and look up words. Search tools can help you find a particular passage. You can also change the text size—an important feature for people with visual difficulties. People will always love printed books, but it's nice to have a whole library in your pocket.

Founding father
The American author Michael Hart created a digital version of the United States Declaration of Independence in 1971. This is often considered the first e-book.

Emoji:
Instant communication

When you talk with someone in person, their facial expressions and tone of voice let you know how they are feeling. But how to capture that in a text? One way is to use *emoji*. 😊 These simple symbols—such as 😲 and 💩 —are everywhere in online communication. It is hard to imagine life online without them!

Emoji have their roots in symbols called *emoticons*. Emoticons are created using standard keystrokes. For example, the popular smiling face ":)" can indicate humor or pleasure.

Emoji are an evolved version of emoticons. The Japanese company NTT DoCoMo introduced the first emoji in 1995. Each image was just 12 by 12 pixels, very simple compared to modern emoji!

There are now thousands of emoji, and new ones are created all the time. There are faces, animals, objects, flags, and symbols. Each emoji has a standard name, such as *thumbs up* 👍 or *face blowing a kiss* 😘 .

Emoji can signal emotion or express a thought without using words. They can also be shared by people who do not speak a common language. Emoji can be combined to suggest new meanings. A *sandwich* emoji paired with a *cookie* emoji (🥪 🍪) could represent lunch. An *airplane* emoji matched with a *luggage* emoji (✈️ 🧳) could mean travel.

Standard symbols

By 2010, emoji had become so popular that a group called the Unicode Consortium added them to its list of standard *character sets* (common sets of symbols used in type). They also work to make emoji more inclusive by adding more options for skin tones, hair colors, gender representations, and disabilities.

Emoji and the arts

Emoji have inspired and appeared in many works of art. In 2009, the American artist Fred Benenson published *Emoji Dick*, a version of Herman Melville's classic novel *Moby-Dick* (1851) written entirely in emoji. In 2016, the Museum of Modern Art in New York City acquired the original set of 176 emoji, created by the Japanese designer Shigetaka Kurita.

Fitness trackers:
Taking steps to better health

Does a step count if nobody counts it? Of course, it does. But you can count your steps—and more—with a fitness tracker. You might even get in shape along the way!

Fitness trackers come in many forms. Some clip to clothing. Others are wearable wristbands or rings. A basic fitness tracker is little more than an electronic *pedometer,* a device that counts steps. Advanced fitness trackers can monitor your pulse, your sleep, and more.

Most fitness trackers include a device called a *3D accelerometer.* This device measures the motion of your body to determine how many steps you take. Data from a 3D accelerometer can be used to track your speed, distance traveled, and changes in elevation.

Some wearable fitness trackers can also monitor your heart rate and the amount of oxygen in the blood. They do this by shining an LED light on the skin. The skin's ability to reflect light changes with the flow of blood.

With a computer or smartphone app that connects to the tracker, users can set goals and track their activity over time. A fitness tracker may even remind someone to get up and move if they have been sitting for a while.

Pedometers have been around since the 1600's. But electronic fitness trackers arrived in 2008, with the Fitbit. The first Fitbit was shaped like a clothespin and clipped to clothing. It had a series of indicator lights that formed a flower shape when the wearer reached 10,000 steps. By 2015, Fitbit was selling more than 20 million fitness trackers a year.

First steps

The Italian inventor Leonardo da Vinci sketched an idea for a pedometer in the 1500's. His mechanical design counted steps using a *pendulum*, a weight that swings as the user walks.

10,000 steps?

The Japanese clockmaker Yamasa introduced an early fitness tracker that they named *Manpo-kei,* Japanese for *10,000 steps.* That number was picked just because it sounds good. Studies have shown that the number of steps necessary for a healthy lifestyle varies. But 10,000 steps remains a common goal.

Flexible touchscreens:
Technology's next flex

Do you ever want to roll up your tablet and put it in your back pocket? With flexible touchscreens, that might be possible. Flexible touchscreens are ultra-thin, bendy, and durable screens for electronic devices.

Most device touchscreens are made of a transparent metallic material called indium-tin oxide. Indium is a soft silver-white metal. When indium is mixed with tin and oxygen, the resulting alloy conducts electricity, but is very brittle.

In 2020, researchers tried a new method to make this material thinner and stronger. They heated the indium-tin alloy to 392 °F (200 °C) until it became a liquid. They then rolled it over the surface to print thin sheets. The result was an ultra-thin flexible screen, 100 times thinner than traditional touchscreens.

These new screens have the same chemical makeup as a standard touchscreen, but a different crystal structure. This allows them to bend. The thin sheets are also more transparent. This means the screens require less back light, saving battery power.

Making foldable touchscreens with liquid metal printing might sound high-tech, but it is as easy as printing a newspaper. The process is cheaper and faster than making traditional touchscreens. The biggest challenge may be that the global supply of indium is limited. Researchers and manufacturers are exploring alternatives, such as carbon nanotubes and fine meshes of gold and silver.

Engineers have yet to develop an entirely rollable, bendable device, so flexible screens today are mostly used on devices that fold in half. But as soon as the rest of the tablet can roll up, so can the screen.

Fold the phone

Chinese manufacturer Royole released the world's first commercially available folding smartphone in 2018, called the FlexPai. The first models were quite expensive.

Plastic to top it off

Flexible touchscreens are covered with a thin film of plastic. The plastic film protects the flexible touchscreen, but it is prone to scratches!

HOVERBOARDS:
Almost like flying

A hoverboard looks a bit like a skateboard, but with two big wheels facing sideways. The hoverboard doesn't really hover above the ground, but rolls itself forward on battery power—no pushing required.

A hoverboard rider stands facing forward and steers by moving their body. The board moves in the direction the user leans. Want to turn? Move one foot so the hoverboard only uses that wheel. Tilting forward or back controls the speed. And how do you stop? If the rider stands straight, not leaning or increasing foot pressure, the board will stop its motor.

Sensors and pressure pads detect how the rider is moving to control the speed and steer. The board also has a gyroscope inside, a device that spins to stay steady. It adjusts the tilt of the board so it doesn't tip.

Hoverboards are powered by a rechargeable battery and have a top speed of about 10 miles (16 kilometers) an hour, though some go faster.

The Chinese American inventor Shane Chen created the hoverboard in 2014. In China they are called *pinghengche*, which means *balancing wheels*. In 2015, the American scooter company Razor bought the licensing rights to Chen's patent. In the following years, hoverboards became a billion-dollar industry.

Patent poachers

When hoverboards became popular, many other companies started making their own imitations. Some copycat hoverboards were made with poor-quality batteries that are prone to overheat and catch fire. The first explosion in the United States happened in November 2014. By 2016, hoverboard sales were banned on Amazon. Thousands of factories closed. Safer models have been introduced since then.

Safety first

Hoverboards were popular gifts in 2015 and 2016. Thousands of eager kids zoomed around their homes—and crashed! To avoid hoverboard injuries, wear a helmet, wrist guards, and knee and elbow pads.

Image filters:
Change your look

You take a selfie. Same old face. But how would you look with puppy ears or tiger stripes? Image filters on photo apps and social media can change photos to make us look funny, ridiculous, or more beautiful, or simply to improve the quality of a photo.

Image filters work in different ways. The simplest kind alters the color of the pixels—the tiny dots that make up a photo. This can make a photo brighter or darker. It can change the color to more red or blue, or it can turn the image black-and-white or sepia for an old-fashioned look.

A more advanced tool, called image warping, allows you to manipulate the shapes in a photo. You can make a face bulge, twist, get narrower, or become wavy. Image-warping apps use mathematical algorithms that stretch or compress different parts of the image to create these effects. You can use it to change the size and shape of your eyes, nose, mouth, or whole head. A slider allows you to control the amount of warp.

Augmented reality

Social media filters, such as Snapchat Lenses, use augmented reality (AR) to add special effects to photos and videos. The app scans your head and shoulders and can then transform you into a cartoon or distorted version of yourself. It can show you wearing different makeup, clothing, jewelry, and other accessories. AR filters also have many other uses (see page 14).

Photo play

In the early days of photography, people sometimes retouched photos using paint, ink, or airbrushing. Occasionally, different images were spliced together. In 1860, a photo splice put the head of U.S. President Abraham Lincoln on the body of another politician, John Calhoun. Photo manipulation became easier with digital photography. Photo-editing programs, such as Adobe Photoshop, can erase red-eye, spots, and wrinkles from faces, and even add or erase people from groups. The first image filters for social media apps arrived around 2010.

Miniaturized cameras:
Smaller and better

The first cameras in the 1830's were bulky contraptions and required toxic chemicals, darkrooms, and special equipment to develop photos. Today's digital cameras can make high-resolution photos instantly, and fit in a pocket—or even in a pill.

Most smartphones today are equipped with a camera. Like all cameras, these collect light through a lens, usually on the back of the phone. Some phones have several lenses for taking different kinds of photos. The phone's internal computer stores the image as digital code. Unlike an old camera, you can view the digital image instantly. You can even edit the photo on your phone and post it to social media.

The GoPro camera is a small video camera designed for filming action. These cameras have wide-angle lenses and waterproof cases, and they can be mounted on automobile dashboards, bikes, and helmets. They are popular with outdoor enthusiasts and other content creators.

Tiny cameras are also helping out in medicine. Surgeons use them to see inside small incisions during surgeries. Gastroenterologists can examine a patient's digestive tract using a pill-sized capsule containing a miniature camera, a light source, and a transmitter. As the capsule travels through the digestive system, the camera snaps thousands of images, which are transmitted wirelessly to a data recorder worn by the patient.

To serve and protect

Body cameras, like the GoPro used by sports enthusiasts, have been used by law enforcement to record public interactions. This footage can be used later in court.

The world's smallest camera is the "Micro Camera" developed in Germany. This camera measures just 0.7 millimeters across—about the size of a grain of salt! The tiny size allows it to film hard-to-reach areas during industrial inspections.

Older than you think

Tiny cameras have been around since the 1800's. The American inventor Robert D. Gray designed a miniature spy camera in 1885. It took photos on glass slides with a lens designed to fit through a buttonhole.

N95 MASK:
Function over fashion

Since the COVID-19 pandemic in 2020, face masks have become a common sight. Made of cloth or paper, they come in many shapes and colors. But one type is far more effective than the others at stopping germs—the N95 mask.

The name N95 tells you that the face mask is made to trap non-oil (N) particles, and that it filters out 95 percent of airborne particles, including very small ones (0.3 microns in diameter) that can contain viruses.

N95 masks are made of plastic fibers thinner than a strand of hair. A machine weaves the threads together, then hot air is used to melt the fibers. Melting the threads creates a dense mesh that allows extremely tiny air molecules to flow through, while stopping particles—including infectious viruses.

N95 masks have long been used by healthcare workers and others at risk of infection. To be effective, the mask must form a complete seal around the nose and mouth, with no gaps. Most masks include a bendable metal nose strip for a tight fit.

Steamy glasses

People who wear glasses usually know if their N95 fits properly. Any gaps between the edge of the mask and your face will make your glasses fog up. This can be avoided by making sure your mask is tight along your nose and cheeks before you put your glasses on.

Some people feel that wearing an N95 mask makes it harder for them to breathe, especially while exercising or for individuals with respiratory problems. But N95 masks do allow plenty of air through and are designed for medical professionals to wear for hours at a time. A mask may take some getting used to, but a little discomfort is better than a long illness.

The real deal

Genuine N95 masks are stamped with N95 and a certificate number. KN95 and KF94 masks are similar to N95 masks, but are certified in China and South Korea, respectively. They provide the same protection as N95 masks. Cloth and surgical masks provide some protection, but they are not as effective as an N95.

Owlet Dream Sock:
Jumping into parenthood feet first

Parents of newborn babies have plenty to keep them awake at night. But some are getting a little help from an unlikely invention—a high-tech baby sock.

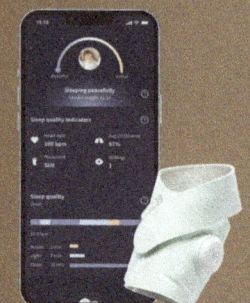

The Owlet Dream Sock monitors a baby's motion and crying. It also records the room temperature and humidity. All this data helps Owlet determine the best time, place, and temperature for naps. Owlet also sells a camera monitor to pair with the sock. It can reassure parents that baby is sleeping, helping them to sleep better as well.

The Dream Sock has its roots in another innovation: the Smart Sock. In 2012, the American engineer Kurt Workman helped care for his aunt's twin babies, who were born prematurely (early). A friend who was a nurse told Workman about a technology called *pulse oximetry*. This shines light on the skin at specific wavelengths to measure heart rate and blood oxygen levels. The technology is used in hospital monitors that clip to a patient's finger.

Workman built a small pulse oximeter into a sock for infants. He added wireless technology to allow the sock to communicate with a monitor or smartphone. He and other partners founded the Owlet company in 2013 to market the invention.

In 2015, Owlet released the Smart Sock for infants. The sock wrapped around a baby's foot and monitored heart rate and blood oxygen levels. These vital signs are a concern to parents because of sudden infant death syndrome (SIDS). The rare condition involves the unexpected death of an otherwise healthy infant.

Over 1 million Smart Socks were sold. In 2019, the U.S. Food and Drug Administration warned that the Smart Sock was being sold as a medical device without proper approval. In 2022, Owlet applied for approval to begin selling the device again.

Infant surveillance

The Dream Sock is just the tip of the iceberg when it comes to keeping an eye on baby. High-tech baby monitors incorporate high-definition cameras, motion sensors, sound detection, and more.

SIDS Education

Sudden infant death syndrome is also called crib death or cot death. Most SIDS deaths occur among infants between 1 and 6 months of age. Scientists have identified several risk factors for SIDS, including the infant's sleeping position, hazardous sleeping conditions, and exposure to cigarette smoke before and after birth. Doctors recommend that healthy babies be placed on their backs for sleep during the first six months of life.

Personal safety alarms:
Small device, loud protection

Nobody likes to think about being attacked by a mugger or other criminal. In general, such attacks are quite rare. But people may feel safer knowing they have some protection. One modern option is the personal safety alarm. These electronic devices are small enough to fit on a keychain. But they can be a big help in an emergency.

A personal safety alarm is designed to emit a loud noise during an attack. The ear-splitting sound may scare off the attacker. It may also alert bystanders that the user needs help. Some also set off flashing lights and call for help over cellular networks.

There are many models of personal safety alarms. Most are designed so they can be triggered with a simple push of a button or pull on a strap.

Before the personal safety alarm, some people carried shrill whistles to blow in the event of an attack. But it can be hard to blow a whistle if you are running or being held down. Personal safety alarms are easier and safer to use than such weapons as pepper spray. You cannot injure yourself with an alarm. It also cannot be used against you. Personal safety alarms are allowed through airport security and in other places weapons cannot go.

The next generation of personal safety alarms will have even more features, allowing users to connect to smartphones and share their location with loved ones.

A safer alternative

Personal safety alarms are safer than carrying weapons or pepper spray. This makes them a popular choice for parents who want to protect their children. Alarms are also safer and easier for the elderly and the disabled.

They're Birdie

The She's Birdie company, founded by the sisters Amy Ferber and Ali Ferber Peters, makes personal safety alarms designed especially for women.

SELF-BALANCING SCOOTER

Want a quick way to zip around town that doesn't take up much space or wear out your sneakers? How about a self-balancing scooter?

A self-balancing scooter is a small platform with either one wheel in the middle or two wheels to the sides. Each wheel is connected to independent electric motors. Other motors are used to keep the platform level. A small computer reads sensors and gyroscopes and controls how the motors run.

Each half of the standing platform can tilt a bit. If the rider slightly leans forward, the scooter starts to move. If the rider leans back, the vehicle brakes. If the rider leans to the right or the left, it turns. A rider simply shifts their body weight and leans in the direction they wish to go.

Segway, a self-balancing scooter invented by American engineer Dean Kamen, debuted in 2001. Segways have two wheels and a tall column in front with handlebars. Instead of relying solely on leaning, users could guide the Segway using the handles. But Segways were expensive and heavy, so they did not catch on for commuting.

A single-wheeled self-balancing scooter (sometimes called onewheel) has a single large wheel, facing either forward or sideways. The first single-wheel scooters were produced in 2007. They are increasingly popular because they are small, light, and portable. They can also handle rough surfaces.

New scoooter models have built-in Bluetooth speakers, cameras, and even internet connections so riders can live-stream their journeys.

Smaller is better

Segway scooters, introduced in the early 2000's, became popular for guided tours and police work but did not catch on as personal transport. Bejing-based Ninebot acquired Segway in 2015. In 2016, they released the Segway miniPRO, a smaller version of the Segway scooter.

Scooter polo

The International Segway Polo Association is a league that uses self-balancing scooters to play polo instead of traditional horses.

57

SELF-DRIVING CARS:
Fast lane to the future

Imagine you have just enjoyed an evening concert in the city. The music was wonderful, but now you are tired. What if instead of a long drive or bus ride home, you could relax in a comfortable, quiet taxi—with no driver? The idea was once futuristic. But self-driving cars and other vehicles are already hitting the streets in a few cities. And they will be traveling to more destinations soon.

Autonomous (self-driving) vehicles have been a dream for years. But they are becoming a reality thanks to fast, powerful computers and better sensor technology.

Lidar is one sensor technology that enables self-driving cars to navigate their surroundings. *Lidar* is short for *light detection and ranging.* A lidar unit rapidly scatters laser light in many directions. When a ray from the laser strikes an object, it bounces back to a detector. The more quickly a reflection returns, the closer an object is.

Lidar units can measure millions of reflections each second. A computer uses these to build a detailed, real-time map of the car's surroundings. Self-driving cars also use radar (radio waves) and cameras to detect possible obstacles.

Safety first

Over 1 million people are killed on the world's roads each year. Many accidents are caused by drunk, drowsy, or distracted driving. Even good drivers may lack the reaction speed to avoid a serious accident. Self-driving cars hold the potential to make driving safer by reducing human driver error.

Tesla takes a different route

All major self-driving car companies use lidar in their cars—except for one. The American manufacturer Tesla has never included lidar in its vehicles. The company also stopped putting radar in its cars in 2022. Instead, it is betting on cameras alone. Its vehicles have traveled millions of miles, giving its artificial intelligence programs plenty of experience processing camera data. But several autonomous driving accidents have led some experts to doubt the company's cameras-only approach.

Self-driving cars are already working the streets as *robotaxis* (robotic taxis). A robotaxi can handle trip after trip, with no need for sleep or breaks. It can drive itself to the depot when it needs to recharge or repair. A human safety observer can monitor several robotaxis remotely.

The companies Waymo and Cruise began operating small robotaxi services in the U.S. cities of San Francisco, California, and Phoenix, Arizona, in the late 2010's. Driverless taxis have been approved for operation in the Chinese cities of Beijing, Shenzhen, and Wuhan.

Autonomous vehicles could also be useful in trucking. Trucks transport tons of goods long distances each day. But driving a truck can be a boring and sometimes dangerous job. Companies are developing self-driving trucks to help over the long haul.

At first, autonomous trucks might drive themselves only over long stretches of highway. A local driver could take over for the more challenging tasks of navigating city streets to deliver the cargo.

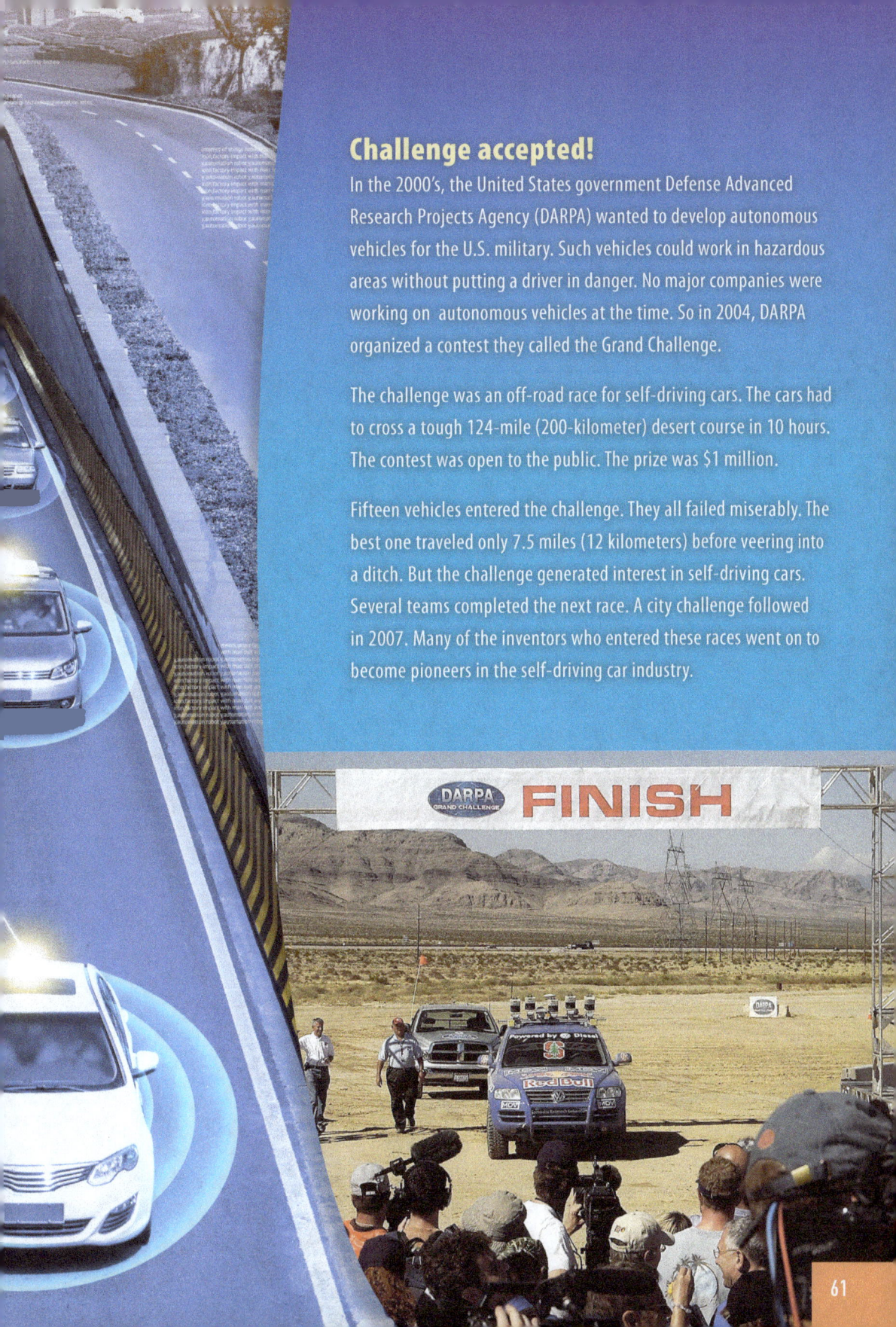

Challenge accepted!

In the 2000's, the United States government Defense Advanced Research Projects Agency (DARPA) wanted to develop autonomous vehicles for the U.S. military. Such vehicles could work in hazardous areas without putting a driver in danger. No major companies were working on autonomous vehicles at the time. So in 2004, DARPA organized a contest they called the Grand Challenge.

The challenge was an off-road race for self-driving cars. The cars had to cross a tough 124-mile (200-kilometer) desert course in 10 hours. The contest was open to the public. The prize was $1 million.

Fifteen vehicles entered the challenge. They all failed miserably. The best one traveled only 7.5 miles (12 kilometers) before veering into a ditch. But the challenge generated interest in self-driving cars. Several teams completed the next race. A city challenge followed in 2007. Many of the inventors who entered these races went on to become pioneers in the self-driving car industry.

SELFIE STICK:

Ready for your closeup?

You're on vacation and want a group photo to post on social media. Do you hand your phone to a stranger and ask them to take your picture? No need if you have a handy selfie stick!

The selfie stick is an incredibly simple device—a collapsible rod with a clip at one end to hold your phone or camera. The selfie stick holds your camera out farther than your arm could on its own, allowing for a wider field of vision, which makes it great for group photos and videos. Some selfie sticks come with a remote trigger mechanism, but you can also just use the camera's delay timer function.

Selfie sticks are a simple invention that solves a common problem—but they also cause some problems. Tourists filming themselves with selfie sticks can run into people, fall down stairs, or damage museum exhibits. Museum curators hate them. Selfie sticks have been banned from many places, including Disneyland, the Tower of London, and the city of Milan, Italy.

The inventor of a selfie-stick prototype, Hiroshi Ueda, first came up with the idea on a family trip in the 1980's. The family asked a child to take their photo, but instead the child ran off with their camera. But Ueda's "extender stick" didn't catch on until smartphones came along and needed a longer reach.

Expensive mistakes

Precious works of art have been damaged or destroyed by selfie-takers reaching for a perfect shot. In 2017, one museum visitor caused over $200,000 in damages when they lost their balance taking a selfie and crashed into an exhibit. Many museums discourage the use of selfie sticks or even ban them.

Long history of selfies

It may seem like "selfies" are a new phenomenon, but people have been taking photos of themselves since photography was invented in the 1800's. Photographers would use mirrors, timers, and remotes to snap their self-portraits.

SMART REFRIGERATORS:
Nerve center of the kitchen

Have you ever stood in line at the grocery store and wondered if you already had milk at home? With a smart refrigerator, you could just ask. Some models can help to plan meals using the ingredients inside. Pretty clever for an icebox!

The first smart refrigerator was released in 2000. It could connect to the internet and would send out an alert if the door was left open. It also let users control the temperature from their phones.

How smart is a smart refrigerator? When it comes to gadgets, *smart* generally means it has computerized features. It can also connect to the internet and mobile devices. Smart refrigerators can network with other smart home appliances. They can also connect to home virtual assistants, such as Amazon's Alexa and Google Home.

A smart refrigerator is a great chef's helper. You can ask it to set a timer or read out the recipe if you forget a step. Instead of trying to scribble on the shopping list with cookie dough on your hands, you can shout at the smart fridge to remind you to buy more sugar.

Some smart refrigerators have screens that can display photos, or act like windows to show you what's in the fridge without opening the door. Others use cameras to provide an interior view. The smart refrigerator can also display your calendar, showing who will be coming over this week. Maybe you should make a double batch of your famous cookies after all!

Smart refrigerators can even help cut down food waste. Optical sensors inside can scan barcodes on packages of meat or milk. The fridge can then track expiration dates and alert you to use ingredients before they go bad. A smart fridge will see the vegetables forgotten in the back. It might also know a good recipe to use them up.

The end of the refrigerator magnet?

Many smart refrigerators have a bulletin board screen. Housemates can use it to display memos or leave notes. You won't need a magnet to post that sweet note to the refrigerator anymore!

Don't lose your cool

For people who like to keep food at precise temperatures, a smart refrigerator will let you change temperature settings on your smartphone.

Smartphones:
The world in your pocket

Not long ago, mobile phones just made calls. Modern smartphones, on the other hand, bring the internet—and the world—to your fingertips. On a pocket-sized smartphone, you can watch videos and sports, play games, record songs and videos, call, text, and much more. But how much do you know about the powerful computer in your pocket?

In 1994, the technology company IBM released a phone called Simon that could send and receive calls, faxes, pages, and emails. It also had a calculator and a calendar.

The term *smartphone* was coined in 1995. But phones with touchscreens, web browsers, and apps did not arrive for a few more years.

The first touchscreen smartphone was the Prada, made by the South Korean company LG in May 2007. Apple introduced its iPhone a month later. These were the first phones with no physical keyboard or stylus (pen). Apple's iPhone could also download *apps* (software applications), for a custom experience.

Nomophobia is the fear of being without your phone. The term comes from *no mobile phone* and *phobia,* meaning fear. Nearly half of the world's population shows signs of nomophobia. They include obsessively checking the phone and not being able to turn it off.

Smartphones and other cell phones connect to networks via radio waves. These waves carry digitally encoded calls, text, and data between the phone and a nearby cell tower. To send and receive signals, every smartphone has an antenna hidden inside.

A smartphone is basically a high-tech pocket computer. It runs software on a central processing unit (CPU), like other computers. Most smartphones use system-on-a-chip (SoC) technology. This incorporates many functions, such as graphics, sound, memory, and WiFi, into a single computer chip.

The most important software in a smartphone is the operating system (OS). This manages all of the smartphone's functions, runs apps, and operates cameras and other sensors. Most smartphones run on the Android OS or Apple's iOS.

As smartphones get smarter, they have largely replaced cameras, video cameras, pagers, telephones, fax machines, timers, and even alarm clocks! Now, if we could only put them down.

Smartphone domination

Surveys suggest that more than 80 percent of people worldwide own a smartphone. That's way more than the number that have a high-school diploma!

SOCIAL MEDIA:
You have a friend request

A notification pings from a smartphone—10 new followers! Everyone likes to make friends. And since the 2000's, more people are connecting with those friends online, through social media.

Social media are websites or apps on computers and smartphones. They let users share news, thoughts, gossip, photos, video, or favorite songs with people in their network, or with the whole world.

Social networking websites started out as a way to connect people with similar interests. Members create a personal profile, telling other users something about themselves. People can look for friends' profiles or strangers who share an interest, and then "add" or "friend" them to connect.

MySpace, launched in 2003, was one of the first big social network sites. In 2004, *Facebook* was created by some students at Harvard University and quickly expanded. By 2008, it had become the most popular social networking site, a position it held for many years.

Founded in 2006, *Twitter* (renamed *X* in 2023) allows social media posts (or *tweets*) limited to 280 or fewer characters. Celebrities and politicians use the platform to post messages to all their followers at once, giving each follower the feeling of personal contact. X and other social media sites use hashtags, the "#" symbol, to allow users to highlight words or trends. That helps users find content they care about.

Going viral

YouTube, TikTok, and other social media sites have given many rising music stars their first shot at fame. Rapper Saweetie got noticed by posting her raps on Instagram. Singer Lil Nas X went viral when his song "Old Town Road" (2018) was featured on TikTok. Social media fame happens when listeners like something and share it with friends, who share it with their friends, who share it with more.

More-social media

Every day at a time decided by the app BeReal, users are notified to post a picture for their friends. The unpredictable time means the pictures show what users are doing right then, with no time for staging. The idea is to share pictures that are more authentic and spontaneous.

69

Instagram is for sharing photos and videos. Advertisers and celebrities often use the platform to post pictures and videos to publicize concerts, movies, and fashion.

Don't know how to do something? There is probably a tutorial on *YouTube* that will show you how, whether you want to learn to crochet or replace a roof. YouTube is a video-sharing website. Videos can be posted by anyone, from DIY hobbyists to movie studios. The website is a popular place to watch music videos, film clips, educational shorts, and live events.

The Chinese company ByteDance introduced *TikTok* in 2016. TikTok videos are short and often informal or silly. Many combine video clips taken by the user with samples of popular songs from TikTok's own library. Rapidly changing trends and challenges are another feature of the TikTok community. In a challenge, users copy a particular video topic or theme—such as a gag, dance, or song—and post their version on TikTok. Sometimes, having music on TikTok can help a musician get famous.

Many businesses and organizations use social media to promote social issues or political activities, or to market their products and services. Social media sites make money by charging advertisers to show their ads to users in between posts.

Influencers are people, companies, or groups that use social media to express their opinions on a topic, review products, or demonstrate an activity, thus influencing their readers or viewers. Influencers may or may not be experts in their field, but they typically have a strong following because people look to them for advice and opinions.

People today spend a lot of time on social media. It's always nice to feel part of a community. But experts warn that no one, especially young people, should measure their self-esteem by how many "likes" they get.

#vscogirl

VSCO is a photo-editing app and website where users post their photos. In 2019, it sparked a "VSCO girl" trend—*#vscogirl*. VSCO girls are young female users who post photos of themselves wearing quirky fashions like puka shell necklaces and oversized t-shirts, as well as trendy brands and products. The VSCO girl has carried over to Instagram, TikTok, and YouTube.

Special interests

"Interest networks" organize members around a topic or goal rather than personal relationships. *LinkedIn* is a popular career site. People use it to find business contacts and search for jobs. *Pinterest* allows users to "pin," or save, links, ideas, and photos they find online to a "board" that can be shared. Boards might be used to collect ideas for wedding themes, recipes, cool toys, or other projects.

TABLET COMPUTERS:
Between a phone and a laptop

Bigger than a smartphone but smaller than a laptop, a tablet computer is sometimes just the right size.

A tablet is a wireless mobile device with a touchscreen display. Tablets vary in size, from around 6 inches (15 centimeters) to 18 inches (46 centimeters). They do not have keyboards. Like other computers, the brain of a tablet is its microprocessor. Tablets typically have smaller processors and less internal memory than laptops. They are powered by a rechargeable battery. Tablets run downloadable programs called apps, similar to a smartphone.

Why tablet?
These devices got their name because they resemble the slate or clay tablets that scribes of ancient civilizations once used to write on.

Tablets have accelerators and gyrometers inside, so they can tell which side of the screen is "up" no matter how you turn it. They can connect to computer networks over Wi-Fi or cellular networks, and have a Bluetooth receiver to connect to other Bluetooth devices, such as an external keyboard or electronic drawing stylus.

Most modern tablets use capacitive touch screens. These are made up of several layers of glass and plastic, coated with a thin layer of copper that conducts electricity. The material responds when touched by another electrical conductor, such as your finger or a stylus.

Tablets with a drawing stylus are especially popular with artists. Drawing apps allow the artist to choose what style the stylus will draw in, so it can look like a pencil, or a paintbrush, or chalk, in any color you like. The result is mess-free digital art.

From Stylator to iPad

Early versions of tablet computers used a stylus to input data to a computer. One was called the Stylator (1957). The first true tablet computers were introduced in 1987. However, they were not commercially successful. The first popular tablet was Apple's iPad, launched in 2010. This began a tablet boom, with other models soon following, such as the Samsung Galaxy Tab, the Motorola Xoom, the HP TouchPad, and the Amazon Kindle Fire.

VIDEO SHARING:
Uploading yourself to stardom

Every day, users share millions of videos on such services as YouTube, Instagram, and TikTok. In a multimedia world, everyone has the chance to see and be seen.

Video sharing became possible in the 2000's with the rise of smartphones and social media. Modern smartphones let anyone record high-quality video. Computer algorithms got better at compressing video files, making it possible to stream or send them. And social media sites provided a place to share videos and friends to share them with.

Today, there are dozens of websites and mobile *apps* (applications) that help people share video content. Some, such as YouTube, TikTok, and Instagram, were built for video. Other social media sites, such as Facebook and Twitter, also support video sharing.

YouTube was founded in 2005 and soon became the most popular site for watching all kinds of videos. Any users can view, upload, and comment on videos. Content creators can set up their own channels, and fans can "subscribe" (for free) to get notified when new videos are posted.

Instagram and Snapchat are often used for more personal sharing. On Snapchat, videos and messages disappear after being viewed.

On TikTok, content creators post short videos set to music. They might do a dance, make a joke, or share funny moments. The app helps users edit videos and add sound and funny effects.

To find videos, users browse a scrolling "feed" of videos posted by friends mixed with videos that the TikTok algorithm recommends (plus ads). If someone sees a video they like, they can share it with friends and followers.

TikTok users can respond to other videos in a duet. In a duet, two videos are posted side by side and run simultaneously. Today's topic: Are we famous yet?

Trends and challenges

TikTok videos are often informal and silly. The site uses labels called *hashtags* (#) to organize similar videos. Hashtags let users participate in trends or challenges. In a challenge, users riff on a specific theme, such as a visual gag, dance, or song. They can post their version on TikTok using the challenge hashtag.

DIY (Do it yourself)!

Video sites are not just fun and games. Many people use them to share important information. On YouTube, you can watch physics lectures, see how proteins fold, or learn how to cook or restore furniture.

VIRTUAL ASSISTANTS:

Handy little helpers

"Hey Siri," … "OK Google," … "Alexa…" If you say these magic words near certain devices, they might wake up the machine's virtual assistant (VA). You can then ask it questions or get it to perform tasks. But what are virtual assistants, and how do they work?

A VA is a type of computer program that can understand and respond to human speech. They can find information on the internet, suggest music or TV shows, tell jokes, manage schedules, send emails, order a taxi or takeout, and control other devices in our homes.

Amazon's Alexa device is a VA, and so is Google Home. Some VA's, such as Apple's Siri, are built into the operating systems of mobile devices. They are also part of mobile apps that help you place orders or buy products.

VA's use artificial intelligence (AI) to interpret our questions and carry out instructions. AI also helps them learn our habits and preferences over time, so they can get better at predicting our needs and making suggestions we are likely to enjoy. Just be careful what you say around that speaker.

Natural language processing

VA's use a type of AI program called natural language processing (NLP) to understand human questions and instructions and match these to commands it can carry out. NLP learns by analyzing millions of voice samples to build a dynamic database of human vocabulary, grammar, and sentence structure. VA's also use natural language understanding (NLU), which helps them work out a user's intentions from the context. Finally, there is natural language generation (NLG), which enables VA's to respond with human-sounding speech.

VA's differ from traditional chatbots, which companies once used to help customers on their websites. Today, most chatbots are powered by AI, but originally they could only provide pre-programmed answers and often struggled to understand customers.

Computer vision

Some VA's have cameras and use a type of AI called computer vision. This enables them to interpret facial expressions and body language and identify objects. By matching a person's spoken words to the movement of their face and mouth, a VA can get better at understanding speech.

Index

accelerometers, 40
adaptive clothing, 4-5
Adobe Photoshop (software), 47
air fryers, 10-11
alarms, 30, 54-55
Alexa (virtual assistant), 64, 76
algorithms, 6, 46, 74-75
Amazon (company), 37, 45, 64, 73, 76
Android OS (operating system), 12, 67
Antheil, George (musician), 19
Apple (company), 12, 66-67, 73, 76
apples, 26-27
apps (software), 12-14, 20, 34, 40, 46-47, 66-69, 71-76
artificial intelligence (AI), 34, 59, 76
augmented reality (AR), 14-15, 47

baby monitors, 52-53
BeReal (social media), 69
bladeless fans, 16-17
Bluetooth, 18-19, 56, 73
ByteDance (company), 70

central processing unit (CPU), 67
charging stations, 20-21, 33-34
chatbots, 77
Chen, Shane (inventor), 44
computer vision, 77
copper, 73
Cosmic Crisp (kind of apple), 26-27
COVID-19 pandemic, 11, 50
crowdfunding, 28-29
Cruise (company), 60

Defense Advanced Research Projects Agency (DARPA), 61
delivery robots, 30-31
domestic robots, 32-33
Dyson, James (inventor), 17

e-ink, 36-37
e-readers, 36-37
electric cars, 20-21
electricity, 16, 20-21, 36, 42, 56, 73
emoji, 38-39
emoticons, 38

Facebook (social media), 28, 68, 74
Fitbit (fitness tracker), 40

fitness trackers, 40-41
flexible touchscreens, 42-43
folding smartphones, 43
Food and Drug Administration (FDA), 52

General Electric Company (GE), 21
Global Positioning System (GPS), 22-23, 30
gold, 42
Google Home (virtual assistant), 64, 76
GoPro cameras, 48-49
Gray, Robert D. (inventor), 49
gyroscopes, 44, 56

Haartson, Jaap (inventor), 18
hashtags, 68, 75
hoverboards, 44-45

IBM (company), 12, 66
image filters, 12, 46-47
indium-tin oxide, 42
influencers, 70
Instagram (social media), 69-71, 74-75
iOS (operating system), 12, 67
iPad (tablet computer), 73
iPhone (smartphone), 66

78

Kamen, Dean (inventor), 56

Lamarr, Hedy (inventor), 19
Leonardo da Vinci (inventor), 41
LG (company), 66
lidar, 58-59
LinkedIn (social media), 71

Micro Camera, 49
microprocessors, 72
miniaturized cameras, 48-49
MySpace (social media), 68

natural language generation (NLG), 77
natural language processing (NLP), 34, 77
natural language understanding (NLU), 77
N95 masks, 50-51
Nokia phones, 13
nomophobia, 66

operating system (OS), 12, 67, 76
Owlet Dream Sock (baby monitoring device), 52

pedometers, 40-41
Pinterest (social media), 71
police body cameras, 49
Prada (smartphone), 66
pulse oximetry, 52

radar, 58-59
radio waves, 18, 58, 67
refrigerators, 26, 64-65
robotaxis, 60

satellites, 22-23
scooters, 56-57
Segway scooters, 56-57
self-driving cars, 58-61
selfie sticks, 62-63
She's Birdie (company), 55
silver, 42
Simon (smartphone), 12, 66
Siri (virtual assistant), 76
Smart Sock (baby monitoring device), 52
smartphones, 12, 14-15, 20, 40, 43, 48, 52, 54, 65-68, 74
Snapchat (social media), 47, 75
social media, 12, 28, 46-48, 62, 68-70, 74
sports analysis, 6-9
Stylator (computer), 73

stylus, 73
sudden infant death syndrome (SIDS), 52-53
system-on-a-chip (SoC) technology, 67

tablet computers, 12, 36, 42, 72-73
Tesla (company), 59
TikTok (social media), 69-71, 74-75
Toshiba (company), 17
touchscreens, 12, 36, 42-43, 66, 72
Twitter (social media), 68, 74

Ueda, Hiroshi (inventor), 62

video sharing, 70, 74-75
virtual assistants (VA's), 64, 76-77
virtual reality (VR), 24-25, 28
VSCO (social media), 71

Waymo (company), 60
Workman, Kurt (inventor), 52

YouTube (social media), 69-71, 74-75

Acknowledgments

Cover   © Yuliya Koshchiy, Shutterstock; © Fahroni/Shutterstock; © Julija Sh, Shutterstock; © Chay_Tee/Shutterstock; © Gorodenkoff/Shutterstock; © Marciobnws/Shutterstock; © Thaspol Sangsee, Shutterstock; © SiljeAO/Shutterstock; © Pablo Lagarto, Shutterstock; © Metamorworks/Shutterstock; © Mix Tape/Shutterstock

4-5   © b.asia/Shutterstock; © NYS/Shutterstock; © Studio KIWI/Shutterstock; © Bezuglova Evgeniia, Shutterstock

6-7   © Michael Kraus, Shutterstock; © Pasko Maksim, Shutterstock

8-9   © vectorfusionart/Shutterstock; © aNorther/Shutterstock; © Maxim Gaigul, Shutterstock; © janews/Shutterstock; © Jurik Peter, Shutterstock

10-11   © GSDesign/Shutterstock; © Marciobnws/Shutterstock; © Twinsterphoto/Shutterstock; © UnderhilStudio/Shutterstock

12-13   © ESB Professional/Shutterstock; © creativerse/Shutterstock; © zaozaa19/Shutterstock; © Kaspars Grinvalds, Shutterstock; © rifatho/Shutterstock

14-15   © Stoyan Yotov, Shutterstock; © Zoran Pajic, Shutterstock; © Dan Thornberg, Shutterstock

16-17   © chuyuss/Shutterstock; © Enjoy the Life/Shutterstock; © TY Lim/Shutterstock

18-19   © Bluetooth; © Boule/Shutterstock; © metamorworks/Shutterstock; © Ivelin Denev, Shutterstock; © Mark Higgins, Shutterstock; © Yanawut.S/Shutterstock; © DestroLove/Shutterstock

20-21   © Summit Art Creations/Shutterstock; © Jackie Niam, Shutterstock; © Petair/Shutterstock; © Smile Fight/Shutterstock; © Fahroni/Shutterstock

22-23   © Gorodenkoff/Shutterstock; © Vectorstock1/Shutterstock; © Sergiy Palamarchuk, Shutterstock; © ANGHI/Shutterstock

24-25   © Frame Stock Footage/Shutterstock; © New Africa/Shutterstock; © N Crittenden/Shutterstock; © Pablo Lagarto, Shutterstock

26-27   © Gorodenkoff/Shutterstock; © EcoPictures/Shutterstock; © Brent Hofacker, Shutterstock; ©2019 Washington State University; © Have a nice day Photo/Shutterstock

28-29   © Da-ga/Shutterstock; © Billion Photos/Shutterstock; © Peppinuzzo/Shutterstock

30-31   © Dabarti CGI/Shutterstock; © Suwin66/Shutterstock; © MONOPOLY919/Shutterstock; © Julija Sh/Shutterstock; © Chad Robertson Media/Shutterstock

32-33   © Studio-N/Shutterstock; © sdf_qwe/Shutterstock; © Studio Nut/Shutterstock; © Takebus6/Shutterstock; © Miriam Doerr Martin Frommherz/Shutterstock

34-35   © rusty12/Shutterstock; © Stock-Asso/Shutterstock; © Lev Radin/Shutterstock; © UlfsFotoart/Shutterstock

36-37   © Ruth Grimes, Alamy Images; © Susan Law Cain, Shutterstock; © TierneyMJ/Shutterstock

38-39   © Gorodenkoff/Shutterstock

40-41   © cunaplus/Shutterstock; © Everett Collection/Shutterstock; © Focus and Blur/Shutterstock; © Mark Rademaker, Shutterstock

42-43   © cobalt88/Shutterstock; © Wit Olszewski/Shutterstock; © Bjoern Wylezich, Shutterstock; © Hussein Satrio, Shutterstock; © ra2 studio/Shutterstock

44-45   © maxbelchenko/Shutterstock; © Yuliya Koshchiy, Shutterstock; © Vasyliuk/Shutterstock; © Melnikov Dmitriy, Shutterstock

46-47   © Oksana Klymenko/Shutterstock; © Yuliya Chsherbakova, Shutterstock; © curiosity/Shutterstock; © Cranach/Shutterstock; © Marketa Kuchynkova, Shutterstock

48-49   © M-Production/Shutterstock; © Kicking Studio/Shutterstock; © Primakov/Shutterstock; © savanterpro/Shutterstock; © Petrovicheva Mariia, Shutterstock; © Marko Aliaksandr, Shutterstock; © Lutsenko Oleksandr, Shutterstock

50-51   © Mike Parolini, Shutterstock; © AlphaTravels/Shutterstock; © JN 999/Shutterstock; © Manu Padilla, Shutterstock

52-53   © New Africa/Shutterstock; © ronstik/Shutterstock; © Anelo/Shutterstock; © Prostock-studio/Shutterstock

54-55   © Prostock-studio/Shutterstock; © SiljeAO/Shutterstock; © moomin201/Shutterstock; © SiljeAO/Shutterstock

56-57   © Aleksandar Todorovic, Shutterstock; © nnattalli/Shutterstock; © ImYanis/Shutterstock; © Caftor/Shutterstock; © Colin Underhill, Alamy Images

58-59   © Metamorworks/Shutterstock; © Flystock/Shutterstock

60-61   © Zuma Press, Inc./Alamy Images; © Zapp2Photo/Shutterstock

62-63   © Ground Picture/Shutterstock; © Jeffrey J Coleman, Shutterstock; © lazyllama/Shutterstock; © Adi purnatama, Shutterstock

64-65   © Michael Ventura, Alamy Images; © Andrey Popov, Shutterstock; © roberaten/Shutterstock

66-67   © ivansnap/Shutterstock; © Olha Yefimova, Shutterstock

68-69   © mundissima/Shutterstock; © antoniodiaz/Shutterstock; © RoBird/Shutterstock; © Christian Bertrand, Shutterstock; © Thaspol Sangsee, Shutterstock

70-71   © XanderSt/Shutterstock; © Antlii/Shutterstock; © metamorworks/Shutterstock; © Jacob Lund, Shutterstock; © GaudiLab/Shutterstock

72-73   © Tero Vesalainen, Shutterstock; © Kamira/Shutterstock; © TippaPatt/Shutterstock

74-75   © Chay Tee, Shutterstock; © Kaspars Grinvalds, Shutterstock; © Prostock-studio/Shutterstock

76-77   © metamorworks/Shutterstock; © whiteMocca/Shutterstock; © Sergey Nivens, Shutterstock; © Wright Studio/Shutterstock